CHANGING CLIMATE

Sally Morgan

FRANKLIN WATTS

NEW YORK • LONDON • SYDNEY

First published in 1999 by
Franklin Watts
96 Leonard Street, London
EC2A 4XD

Franklin Watts Australia
14 Mars Road
Lane Cove
NSW 2066

EARTH WATCH: CHANGING CLIMATE was produced for
Franklin Watts by Bender Richardson White.
Project Editor: Lionel Bender
Text Editor: Jenny Vaughan
Designer: Ben White
Picture Researchers: Cathy Stastny and Daniela Marceddu
Media Conversion and Make-up: MW Graphics, and Clare Oliver
Cover Make-up: Mike Pilley, Pelican Graphics
Production Controller: Kim Richardson

For Franklin Watts:
Series Editor: Sarah Snashall
Art Director: Robert Walster
Cover Design: Jason Anscomb

A CIP catalogue record for this book is available from the British
Library.

ISBN 0-7496-3302-6 (hbk)   0-7496-3593-2 (pbk)

Dewey classification 551.6

Printed at Oriental Press, Dubai, U.A.E.

Picture Credits
Oxford Scientific Films: cover main photo & pages 22-23 top (Richard
Packwood), cover small photo & pages 6 (Daniel J. Cox), 5 top (Martyn
Chillmaid), 13 top (Michael Fogden), 19 top (Edward Parker), 20-21
(Kim Westerskov), 27 bottom (William Paton/Survival Anglia). **The
Stock Market Photo Agency Inc.:** cover globe & pages 1, 6-7, 9, 11
bottom, 16, 18, 22-23 (J. M. Roberts), 24. **Science Photo Library,
London:** pages 4 (Alex Bartel), 8 (European Space Agency), 12, 14 (John
Eastcott & Yva Momatiuk), 17 top (Hank Morgan), 17 bottom (Carlos
Munoz-Yague/Eurelios). **Environmental Images:** pages 5 bottom (Rob
Visser), 26 (Martyn Bond), 28 (Robert Brook). **Ecoscene:** pages 10 (Kay
Hart), 11 top (Jim Winkley), 15 (Miessler), 19 bottom (Chris Knapton),
21 left and right, 27 top (Glover), 29 top (Angela Hampton).
e.t. archive: page 13 (National Gallery, London). **Panos Pictures:** page
23 (Jeremy Homer), 25 top (Fred Hoogervorst), 25 bottom (Sean
Sprague), 29 (J. Holmes).

Artwork: Raymond Turvey

CONTENTS

CHANGING CLIMATE

In the last few years, the climate has been in the news almost every day. We read about heat waves, gales, droughts, ice storms and floods. The Earth is getting warmer, and this is making the world's climate change. Scientists call this increase in temperature 'global warming'.

Holidaymakers in Spain enjoy the sunshine and sea. Climate changes are warming the land, sea and air.

Why is the climate getting warmer?

The world's climate has changed naturally many times in the past: there have been ice ages, and there have been times when the planet has warmed up very quickly. Some scientists believe that today's climate changes are also happening naturally.

But many scientists think that the changes are happening because of human actions. For example, we are burning more fossil fuels – coal, oil and natural gas – than ever before. This releases waste gases into the air in ever-increasing amounts.

An almost dry reservoir in Britain. In recent years, Britain's usually wet climate has been considerably drier.

What will happen?

Nobody knows exactly how our climate will change. Some places may get drier and have year-round temperatures up to 4°C hotter. Other places may become several degrees cooler. Stormy weather may become more common. Glaciers and icebergs may start to melt and never form again.

Whatever happens, climate change will affect people as well as the natural world. But there are things that we can do to slow down the changes and to minimize any ill-effects.

Land flooded as a result of the Mississippi River in the United States becoming overfilled with rainwater.

WHAT IS CLIMATE?

Climate is the average weather in a place over a long time. For example, Europe and North America have a climate with warm summers and cold winters. The weather, however, can change from day to day. It might be sunny one day, and rainy the next.

Different climates

In the tropics (the regions near the middle of the Earth), the climate is warm all the year round. Further north and south, are the temperate regions. These have a climate of warm summers and cold winters.

The climate in polar regions, near the North and South Poles, is very different. Summers are short and the temperatures hardly rise above freezing (0°C). Winters are long and dark as the Sun does not appear above the horizon.

Polar bears spend all their lives in an Arctic climate.

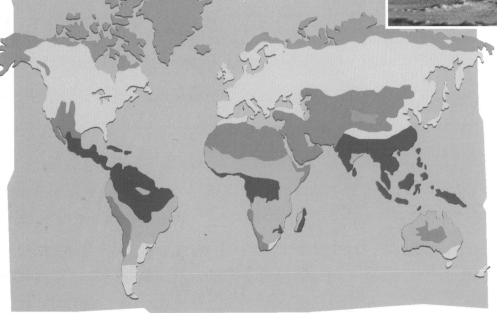

WORLD CLIMATES

- Polar
- Temperate
- Desert
- Tropical
- Equatorial

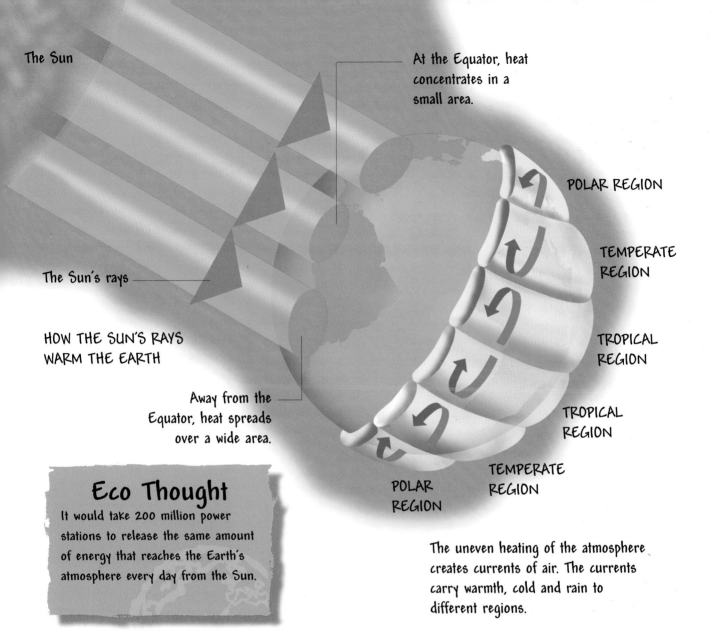

The Sun

The Sun's rays

HOW THE SUN'S RAYS
WARM THE EARTH

At the Equator, heat
concentrates in a
small area.

Away from the
Equator, heat spreads
over a wide area.

POLAR REGION

TEMPERATE
REGION

TROPICAL
REGION

TROPICAL
REGION

TEMPERATE
REGION

POLAR
REGION

The uneven heating of the atmosphere
creates currents of air. The currents
carry warmth, cold and rain to
different regions.

How the Sun affects climate

Different parts of the Earth's surface receive
different amounts of the Sun's heat. This is what
creates different types of climate. Over the
tropics, the Sun is high in the sky. Its rays of heat
and light beam straight down to the Earth. Each
ray warms up a small area with its full force.

Near the poles, the Sun is lower in the sky.
Its rays travel through more air to reach the
ground and become weaker. The rays hit the
ground at an angle, so their heat is spread over a
wide area.

The lower the Sun is in the sky, the
less heat reaches the Earth's surface.

CLIMATE AND OCEANS

Climate is very complicated. A little less rainfall than usual in a place can affect the temperature, the winds and the amount of moisture in the air. The oceans have a huge effect on climate, just like the Sun.

The patterns of the clouds covering the Earth show which way the winds blow.

Oceans and climates

In the tropics, the Sun's heat warms the top layer of the oceans. This warm water is less dense (heavy) than cold water, so it floats on top of it. At the North and South Poles, the surface water cools, becomes heavier and sinks. All this creates movements in the oceans, called currents.

Currents affect the climate. Warm ocean currents heat the winds that blow on to land and make the local climate warmer. Cold ocean currents cool the winds and the local climate.

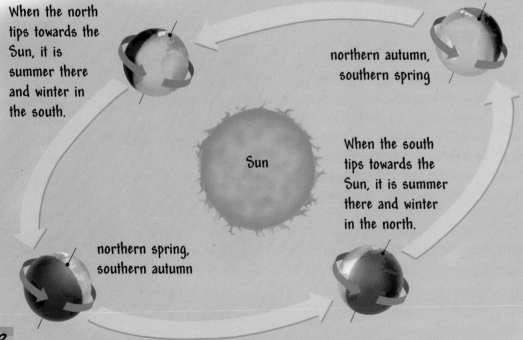

When the north tips towards the Sun, it is summer there and winter in the south.

northern autumn, southern spring

Sun

When the south tips towards the Sun, it is summer there and winter in the north.

northern spring, southern autumn

As the Earth circles the Sun, so the northern and southern halves get unequal amounts of heat during the year. This creates the seasons. Climate changes caused by ocean currents affect how hot or cold these seasons will be.

Changing currents

As the oceans' currents change direction, the way heat spreads around the world changes too. For example, there is a warm current off the east coast of North America that flows close to north-western Europe. This current is called the Gulf Stream. It keeps countries like Britain ice-free in winter. If this current changed, all of north-western Europe would have a much colder climate than now.

This map shows the temperatures of the oceans and continents – the warmest areas are coloured red and the coldest are coloured blue.

Every three or four years, currents in the Pacific Ocean change direction for about nine months. A cold current flowing north along the coast of South America becomes a warm current moving south. This change is called 'El Niño' (Spanish for 'the child'). It affects the weather all around the Pacific. During the period of the El Niño there are often droughts and unusually heavy rains.

Eco Thought

In 1997, there was an El Niño that had a stronger effect than any other in the last hundred years. The temperature of the sea rose so much that it caused serious droughts in Australia and Indonesia. There were forest fires in Borneo, and very heavy rains fell on California and Japan.

Bangladesh in Asia has a hot, wet climate. Unusually heavy rains often cause flooding, as here.

CLIMATE CLUES

We know that the climate is changing because people have written about it for hundreds of years. They recorded the weather conditions, and wrote about storms and floods. Historical records and diaries help us compare the weather of the past with that of today. There are other clues, too.

Clues from ice

Glaciers are masses of ice found at the North and South Poles and in mountains. Ice deep inside a glacier may be thousands of years old. Bubbles of air are trapped in this ice. Scientists can learn about past climates by studying the gases in these bubbles. The more carbon dioxide they contain, the warmer the climate was when they formed.

These glaciers in Alaska in the United States are evidence of thousands of years of a constantly cold local climate.

Tree records

Each year, a tree grows a new ring of wood inside its trunk. In warm, wet years, trees produce wider rings than usual. Scientists remove tiny cores of wood from trees and examine the rings. Trees such as giant sequoias are so old that the rings give clues about the climate over a thousand years ago. In Siberia, in Russia, rings show that the trees grew most in the 1980s and 1990s.

As the Earth's climate gets warmer, spring flowers like these bloom earlier in the year.

This slice through a tree trunk shows the growth rings of wood. The dark band round the edge of the trunk is the tree bark.

On the Ground

Near Bryce Canyon in Utah, United States, there are some of the oldest trees in the world. These are bristlecone pines, and the oldest are 5,000 years old. They among the best clues we have about past climates.

Taking Part

Have a look at annual rings on a tree stump like those in the photograph here. Try working out the age of your tree by counting the rings. The oldest rings are at the centre. In which year did the tree grow the most?

NATURAL CHANGES

Many natural events can cause the Earth's climate to change. For example, the activity of the Sun itself, or erupting volcanoes, can affect the climate.

Volcanic dust

When a volcano erupts, it throws tonnes of dust and droplets of an acid called sulphuric acid into the air. The dust and droplets stop some of the sunlight from reaching the Earth. This cools the Earth down.

Activity of the Sun

The Sun is a huge fiery ball in space. It throws out heat at varying levels. It often gets hotter than usual for a week or two, then cools down. The changes in Sun's activity affect the heating of the Earth's surface.

On the Ground

During the Little Ice Age in the 1600s, the River Thames froze over and the people of London held fairs, called Frost Fairs, on the ice. Icebergs were seen off the coast of Norway. Neither of these things happen today.

This is a close-up of the Sun's surface. Its bursts of energy change the Earth's climate.

Past climates

When dinosaurs existed, between 240 and 65 million years ago, climates all over the world were different from today. For example, forests covered parts of Europe and Antarctica.

During the last million years, there have been four periods when the Earth's average temperature was several degrees Centigrade colder than now. Glaciers covered much of the land. These cold periods are called ice ages.

A painting of people in the Netherlands in 1609 walking and skating on a river that has frozen over during the Little Ice Age.

Eco Thought

In the 1930s a series of hot and dry summers caused a drought in large areas of mid-western North America. Crops failed and winds blew the soil away. The area became known as the Dust Bowl.

Millions of years ago, this desert in southern Africa was a swamp and a dinosaur made footprints in the mud.

IN THE GREENHOUSE

On a hot day, sunlight pours through the glass in a greenhouse. Some of it warms the air inside. The glass stops the heat escaping, so the temperature inside increases. There are gases in the Earth's atmosphere that act like the glass. We call them greenhouse gases.

Clouds float above the Earth, where the air is thinner and colder. The atmosphere holds heat close to the Earth's surface.

Global warming

Greenhouse gases keep the atmosphere warm. Without them, the Earth would be too cold for life to survive.

Over the last 200 years, the amount of greenhouse gases has increased. This is like making the glass in a greenhouse thicker, so less heat escapes. Most scientists believe that a build-up of greenhouse gases is what is making the world's temperature rise. Even if there are other reasons for the rise, the increase in these gases could be making global warming worse.

Eco Thought

The most important greenhouse gases are carbon dioxide, methane, nitrous oxide and substances called CFCs. There is nearly a third more carbon dioxide in the atmosphere today than there was 200 years ago.

Why are these gases increasing?

One major cause of the increase in greenhouse gases is that we are burning more and more fossil fuels, such as coal and oil. This produces carbon dioxide, one of the main greenhouse gases. Another cause is the cutting down of forests. Living trees take up carbon dioxide from the air. With fewer trees, more of this gas stays in the air.

A greenhouse traps the Sun's heat, and keeps the plants inside warm, even when the weather outside is cold.

Energy from the Sun pours on to Earth.

Some sunlight is reflected straight back into space.

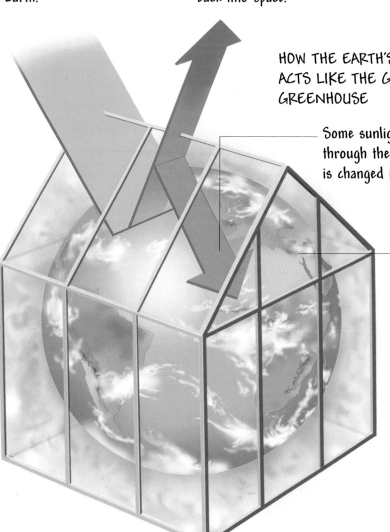

HOW THE EARTH'S ATMOSPHERE ACTS LIKE THE GLASS OF A GREENHOUSE

Some sunlight passes through the atmosphere and is changed into heat energy.

The atmosphere prevents the heat leaving again, so the Earth warms up.

Taking Part

On a sunny morning, put a thermometer in a greenhouse. Make sure the windows are closed. At midday, see how much the temperature has increased. This is what is happening to the Earth, but on a much greater scale.

GETTING WARMER

Around 150 years ago, people started to keep records of the climate. Over this period, ten out of the eleven warmest years on record have been since 1980.

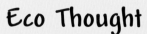

Eco Thought

The global temperature may rise by between 1.5°C and 4.5°C over the next 100 years. If this happens, it will be the biggest increase since the end of the last ice age, over 20,000 years ago.

Keeping track

Today, weather stations, ships and satellites keep track of climate change. They gather information about air and sea temperatures, wind speeds and cloud cover. They feed the information into powerful computers, which have programs that predict how the climates will change.

Although climates have changed in the past, today's changes are distinctly different. Temperatures have never before changed so quickly over such a short period of time. There are more floods, heatwaves and droughts than only 10 years ago.

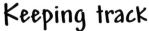
As global warming increases, more river beds will dry up, like this one in Africa.

A scientist records the carbon dioxide level in a sample of the atmosphere. Comparing the level with carbon dioxide levels in the past gives clues about climate change.

Warmer nights

The nights are getting warmer faster than the days, leading to less of a difference between daytime and nighttime temperatures. In many places in the Northern Hemisphere (the northern half of the world) there are fewer frosts at night. This is because greenhouse gases are good at trapping heat at night. As a result, in places spring and autumn climates are very similar to each other.

Scientists use computers to study weather patterns all over the world.

On the Ground

Scientists are carrying out a noisy experiment in the Indian Ocean. Sound travels faster through warmer water. To test if the water is getting warmer, scientists produce a loud, booming noise under water and record the time the noise takes to travel between two places.

GREENHOUSE GASES

There are several greenhouse gases. Some are produced naturally. Others are produced by factories, vehicles and many kinds of industry.

Carbon dioxide

The greenhouse gas carbon dioxide is produced by most living things. They breathe in oxygen, and the cells in their bodies use it as they break down food. Then they breathe out carbon dioxide as a waste product.

Carbon dioxide is also released when anything containing the chemical carbon is burned, for example wood, oil, gas and coal. Power stations, factories and vehicles that burn these fuels produce lots of carbon dioxide. So do burning forests. Each year, more than six billion tonnes of carbon dioxide is released into the atmosphere.

Coal-fired power stations, like this one in Britain, release the greenhouse gas carbon dioxide into the atmosphere.

As the number of humans on the planet increases, so does the amount of carbon dioxide that we produce.

Taking Part

You can monitor the climate. Keep track of weather reports on the radio or television, or from newspapers. See if you can find out if the pattern of weather that you have recorded is different from previous years.

Other greenhouse gases

Methane gas is produced by tiny living things called bacteria which are found in marshy places. It is also released from rubbish dumps, sewage works, termites' nests and cows' stomachs. There is less methane in the air than carbon dioxide, but it is a more powerful greenhouse gas.

Nitrous oxide is a greenhouse gas released by fertilizers and by burning certain fuels. CFCs, which are released from old types of aerosols, packaging and refrigerators when they are destroyed, are also greenhouse gases.

Eco Thought

Methane is a greenhouse gas made up of carbon dioxide and hydrogen gas. A single cow belches 500 litres of methane every day.

Car engines burn petrol, producing several greenhouse gases.

THE RISING SEA

Global warming is making sea levels rise. In the last century, they have risen by 10 to 25 centimetres. Scientists estimate that they will rise another 75 centimetres in the next 50 years.

A rise in sea levels may produce bigger waves that will damage beaches.

Melting ice

When water gets warmer, it expands (takes up more space). When seas warm up due to global warming, the water expands and sea levels rise. Warmer temperatures also make ice melt. The ice sheets that cover Antarctica are breaking up and melting. Glaciers on mountains are getting smaller. The extra water from the melting ice enters the seas.

Islands under threat

Huge numbers of people live on low-lying coastal plains and small islands. As sea levels rise, these places are threatened with floods. Cities such as Alexandria in Egypt, Venice in Italy and Miami in the United States could disappear under water. Large areas of the Netherlands, Belgium and Luxembourg – the 'Low Countries' – would also be flooded.

Stormy weather

Tropical hurricanes are powerful storms that form over warm water and gather heat energy from it. As the oceans get warmer, the storms can gather more heat energy, and become more violent. In future, hurricanes may be much more common – and much more powerful.

These two photographs show the same beach in the Indian Ocean islands of the Maldives. The photo on the left was taken in 1990. Five years later, rising sea levels have washed much of the beach away.

MORE DESERT

Global warming is making deserts expand. Some of the most important crop-growing areas of the world are in danger of being swallowed up by desert.

Drought in Africa, forest fires in Europe

South of the Sahara Desert in Africa, there is an area called the Sahel. Since the 1960s, the average rainfall there and in other parts of Africa has fallen. There have been many droughts. Crops have failed, and in some places, thousands of people have starved. Although other countries have tried to help with food and money, nothing can bring more rain.

The Mediterranean region is north of the Sahara Desert. Summers there are becoming much warmer and drier. This has caused huge forest fires, especially in southern Spain and France, and rivers and reservoirs have dried up.

This water hole in Africa's Sahel region can no longer provide water for cattle.

Less land for crops

If areas of hot and dry climate spread further northwards and southwards, farming will change. Places where sunflowers and maize grow now, will become too dry to grow any crops. Maize may grow in cooler places instead, where wheat grows now. But, in the end, there will be fewer crops – and less food for people to eat.

Eco Thought

If global warming does not slow down, wheat production will drop by a sixth in China and by more than half in India by the year 2010.

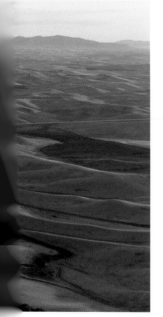

This farmland in the midwest of the United States could once again become barren if global warming continues.

Sunflowers can now be grown in climates that used to be too cool.

PLAGUES OF PESTS

Animals and plants are affected by climate changes, too. A warmer climate may bring new, unwanted animals into a region and they might affect our crops and health.

A scientist studies wheat plants to learn how resistant they are to plant diseases carried by insects.

Carriers of disease

Female mosquitoes bite people and suck their blood. Some of them carry tiny creatures that get into the blood and cause the disease malaria. Others carry viruses (germs) that cause yellow fever and dengue fever.

Climate change affects where mosquitoes are found. Two thousand years ago, southern Europe was warmer and wetter, with swamps where malaria-carrying mosquitoes lived. The climate became drier and cooler. The swamps disappeared, and so did the mosquitoes.

Eco Thought
In South-east Asia, traditional healers use about 6,500 types of tropical plant to treat illnesses that include malaria and stomach ulcers.

A Kenyan farmer sprays his cattle to protect them from insects that carry disease. Global warming may make these diseases more common.

Pests on the move

If the climate warms up again, malaria-carrying mosquitoes may move into holiday areas such as the Mediterranean and Florida. In fact, diseases caused by mosquitoes are already spreading. For example, the highlands of central Africa used to be free of malaria. Now it is turning up there.

Yellow fever has broken out in Ethiopia, which has not happened before. Dengue fever is moving north to Costa Rica, Colombia, Mexico and Texas. If the climate heats up, other insect pests may spread too. Plant pests such as locusts and budworms could destroy more plants and crops.

In tropical Africa and Asia, swarms of locusts often destroy crops. Climate change may help these insects spread further northwards.

On the Ground

Swarms of caterpillars called budworms are eating the needles of the spruce trees in the conifer forests of Alaska. Scientists blame the warmer Alaskan summers for plagues of insects like this. Helicopters are used to spray pesticides over the forests in a battle to save the trees.

FINDING NEW FUELS

Fossil fuels formed from the remains of plants and animals that lived millions of years ago. All fossil fuels contain carbon, which has been locked up in them since they formed. When they are burned, the fuels release carbon dioxide into the air.

These solar cells in California change energy from the Sun (solar energy) into electricity.

Sun, wind and water

We would produce less carbon dioxide if we burned less fossil fuel. We can use renewable sources of energy to generate electricity. These are sources that never run out, such as the Sun, the wind or water. The energy of the wind or of moving water can be used to power generators and make electricity. Solar panels trap the Sun's energy and convert it to heat or electricity, or even power cars.

Eco Thought

Between them, all of the refrigerators in the United States use an amount of electricity equal to the quantity made by 20 large power stations.

Wind turns the blades of these turbines in California. The turbines are connected to a generator that creates electricity.

Waste for energy

Burning fossil fuels adds to the carbon dioxide in the air. Burning plant materials that have grown recently does not. This is because, while they were growing, the plants took carbon dioxide from the atmosphere. Burning them just puts the same amount of the gas back in.

Rotted plant waste can be made into pellets (little blocks) that can be burned without adding to the carbon dioxide levels, too. In many countries, people put waste into pits dug in the ground. As it rots, it releases a gas called biogas that can be used for cooking and heating.

These people in India are making fuel blocks from cow dung. This fuel damages the atmosphere less than fossil fuels do.

On the Ground

There are 2,000 biogas plants in Bangladesh. They work so well that dung from six cows or four buffaloes can make enough biogas for a family of eight people to use for lighting and cooking.

WHAT CAN WE DO?

There are plenty of things we can do to limit the changes people are making to the climate. Some are things each of us can do. Others are for families. There are some things whole schools can do to help. If everybody works together, it can have a big effect.

Save energy

One way to cut the amount of carbon dioxide in the air is to save energy so that less fossil fuels are burned. We can turn down central heating and turn off lights in empty rooms. We can use low-energy light bulbs, which use a fifth as much electricity as ordinary bulbs. New designs of power stations get more energy from fuel, and new kinds of cars use less petrol.

This waste land, left after a factory was demolished, is now used for growing trees which absorb greenhouse gases.

Eco Thought
A typical home produces twice as much carbon dioxide in a year as a car. So, saving energy in the home can help to prevent global warming.

By taking glass, tin and aluminium containers to recycling centres, we help save the energy that would be used making new materials.

Planting trees

Trees absorb carbon dioxide from the atmosphere to make their own food. If we burn wood from specially planted trees, we do not damage existing forests. We can burn the wood without adding to carbon dioxide levels.

We can plant new trees to help absorb carbon dioxide. Fast-growing trees such as willow and birch are ideal for growing for wood for fuel.

Cycling or walking to school or work cuts down on the amount of vehicle fuels that are burned.

FACT FILE

Lots of gas
Although the United States has only four per cent of the world's population, it produces almost twenty five per cent of all greenhouse gases.

Melting ice
Siberia, in northern Russia, is warming up. Once the ground was frozen all the time. Now it is soggy in some areas, and buildings can no longer stand up straight.

Less frost
In the 1950s, in the north-eastern United States, the season when there was frost lasted 11 days longer than it does now.

More hurricanes
Until the 1980s, there had only been one violent cyclone in the last 100 years in the Pacific islands of Western Samoa. In the 1980s, there were three in four years.

Forest fires
There is a growing risk of forest fires in Yellowstone National Park in the United States. Summer temperatures are getting higher and droughts are lasting longer.

Pest invasions
In New Orleans, United States, there was an explosion in the numbers of termites, mosquitoes and cockroaches in 1995. Normally, these insects are killed by winter frosts, but there had not been any frosts for five years.

Too hot for health
In Europe and the United States, doctors estimate that several thousand people died of heart attacks and respiratory (breathing) diseases caused by heat waves in 1995. A July heat wave killed 500 people in Chicago that year.

Rats and disease
Record high temperatures and heavy monsoon rains caused epidemics of rats in India during 1994. The rats spread a disease called bubonic plague, which can be deadly.

Using less energy
If one per cent of American homes replaced their light bulbs with the energy-saving types, in just one year the United States would save: $800 million in electricity costs; two and a half billion kilograms of coal; eight billion kilograms of carbon dioxide and other polluting gases—enough energy to light 300,000 homes.

GLOSSARY

Atmosphere The layer of gases around the Earth.

Axis An imaginary line through the centre of an object, which it spins around.

Biogas A gas given off by plant and animal wastes when they rot down in an underground pit or container.

CFC Chlorofluorocarbon, a gas once used in refrigerators and aerosols, or spray cans.

Climate The usual pattern of weather that is the same over a long period of time.

Conifer An evergreen tree that has needle-like leaves and bears cones instead of flowers.

Current A flow of water or air.

Cyclone A wind blowing in a circle. Some cyclones are very violent.

Drought A long period without any rain.

Equator An imaginary line all the way round the middle of the Earth.

Fertilizer Nutrients added to the soil to help plant growth.

Fossil fuel Coal, oil and natural gas which are formed from plant and animal remains trapped in rocks.

Glacier A slow-moving mass of ice formed in mountains, which creeps down valleys.

Hurricane A powerful tropical storm bringing strong winds and heavy rain. It is also called a cyclone or typhoon.

Malaria A dangerous disease caused by a tiny organism that is carried by the female mosquito.

Northern Hemisphere The northern half of the world.

Pesticide Substance that will kill insect pests such as greenfly and locusts.

Plague Large numbers of a pest affecting a wide area.

Polar From near the North Pole or South Pole.

Renewable Something that can be replaced or regrown, such as trees, or a source of energy that never runs out, such as the Sun or wind.

Reservoir An artificial lake which stores a large quantity of water for cities.

Temperate climate A climate of mild seasons, with warm summers and colder winters.

Tropical The hot, wet regions of the world near the Equator.

Tropics The region of the world, stretching between two imaginary lines around the world. These are the Tropic of Cancer, north of the Equator, and the Tropic of Capricorn, south of it.

Virus Tiny organism that can cause disease.

Volcano A gap in the Earth's crust where hot rocks, lava and gases escape onto the Earth's surface.

Weather The air conditions experienced in one place at one time.

INDEX